MW00453570
Who can the man be?
I spell M-A-T-T.
Matt is a cop on the block.

I am ten.

I am Matt's best friend.

A black cat stops at the top.
I ask Matt to help him.

The men help Matt.

Can Matt get the cat from the top?

Matt gets the black cat.

He hands the cat to me.

I pat the cat's back.

The cat lets me pat him.

I get set to kiss the cat.
Smack!

The cat rests.

Matt and I are glad.